Prologue

"3...2...1...and liftoff, liftoff of Orion 8 heading off to Mars for the first manned landing!" All of us were excited. As the commander of the backup crew for Orion 8 and the official commander of Orion 9, It was joyous to see Orion 8 heading off to it's grand voyage off to a planet unknown.

"Orion, go for throttling down," Houston said. Orion 8 began to throttle down it's main engines. It had to do this or it would break up from the pressure at the altitude in the atmosphere they were at. This call was usually called at T+ 35 seconds on Orion flights.

"Orion, go at throttle up."

"Roger, go at throttle up." The call to have Orion 8's main engines go back up to full throttle was called. They had made it through the hardest part of ascent. They were on there way to orbit.

"Houston, we've had an electricity probl..." As I (Matthew Psy) heard the cut off of the call from Orion to Houston, I turned to the rocket. Or what use to be the rocket. It was now a smoke cloud of charred mess. The external SRB's (solid rocket boosters) were flying off in all directions all over the place. We lost all contact with Orion 8 as well. We saw the command pod plunging toward the ground at over 500 meters per second. The parachutes never deployed, causing them to plunge toward the ground with no hope of escape. I felt lucky, but depressed at the same time, knowing that if something happened to the commander, Martin Brown, then I would've been in that pod, hurling with no escape.

The fact that this happened on the flight before mine had me nervous. What if the same thing happened to my flight, Orion 9?

"Well, It's 4 month's to my flight. I guess we'll find out." My landing pilot, Sean Argon, felt the same way as me. Same as my pilot for Orion 9, Reed Jupiter.

Chapter 1

As the commander of the upcoming flight, Orion 9, I had to make sure everything was fine and ready. If it wasn't good to go, then we would have a bad launch, like Orion 8, 4 months earlier. The flight plan was to rendezvous with the lander that was meant for Orion 8, burn out to Mars, fly through the thin atmosphere and circularize our orbit, perform entry, land, and head home. We all know that plan's aren't meant to work though! So we just had to hope that all would go well.

  The vehicle would carry 3 men to space, 2 to the surface, and close to 7.8 tons to space. We would perform experiments in orbit around the Earth, Sun, and Mars. The scoopers would grab Mars soil and would transfer it into the lander which we would be in. This would allow us to perform more experiments than any rover could do alone. I thought about these things as I sat in the Reentry simulator in NASA's manned space flight facility. I was brought back to my senses from the buzzing of an alarm and the red flash that came with it.

  "Woah, what's the story here? I can't yaw at all!" exclaimed my MEM (Martian Excursion Module) pilot, Sean Argon.
"Must have a thruster problem," I said. "The RCS (Reaction Control System) must have had a malfunction."
"Ah, I see. Look's as if the glitch is making us come too shallow through the Martian atmosphere," Sean said as the reentry warning's continued to beep at us.
"Here, I'm going to take control. Switch me to manual and turn on the SAS system," I ordered.
"Roger, let's not mess this up."

We swapped, and I went to manual control on reentry. I fired up the engine, and put ourselves on a full no going back landing. Thankfully I got us through and we had our simulated deployment of the parachutes and the final touchdown. We had to make this a special one. We got down to the surface, I'll be it using much more fuel than intended.

"Alrighty dirty dog, you ready to ask NASA for our early EVA and go/no go for staying?" I asked sarcastically.
"Sure, why not!" Sean laughed.

We stepped outside the lander sim. and got a congrats for performing the landing with a shutdown RCS system. We were ready for our landing on Mars.

After drilling us with preparation for landing, they decided to put us out there for the rendezvous and docking test. Here in was the hardest part in orbit around the Earth. We had to get ready for our launch at EXACTLY the right time. If we didn't we'd be far off from our target.

"Let's shut some thrusters down on them," a flight tester shut off our thruster blocks on our left and right. This made the craft not able to translate up or down. The pilot, Reed Jupiter, had to be able to dock the vehicle without these, should they fail.

"Woah, wait a sec. I can't translate up or down," Reed said.

"Copy, um, try and get the craft to relocate. We need to get stabilized again. We're drifting far off course," I said.

"Yeah, I see what you mean."

"Rotate the craft around. We can still be able to translate up and down, we just have to rotate around and thrust up and down."

"Copy, I'll do that." Reed went ahead and rotated the vehicle and thrusted into position. He then rotated it back once our target velocity was 0.0 m/s. Once it was reached, we spun back into position and thrusted forward.

"Capture!" Reed exclaimed.

"Thats it! Hah, thats what we've got here! Nice job on that recovery Reed." I said

"Thanks! That was 2 minutes of split terror."

We stepped out of the simulator and finished up the rest of the business. I ran through the list of flight simulations we had done. We had done ascent, docking, landing, reentry, and burn to Mars and back. We were prepared for Orion 9. We just had to wait for 1 more day.
"This is my 5th time in space already!" I thought to myself."Can't wait to return finally."

As my crew and I stepped out of the building, we were greeted by the familiar site of the Orion rocket. It's a wonderful and towering site to see. It's amazing to think that this was the first rocket of it's kind leaving low Mars orbit.

"You guys prepared for the big day tomorrow?" I asked. We all just stood there, amazed.

Chapter 2

When you board a huge pile of fuel with an engine attached, you get pretty nervous. Your body is a-shaking, you're pretty excited, and all in all, you feel pretty good. Last night was pretty good, and we woke up to find ourselves in NASA's dressing and last minute flight prep room. This is where we would get our orbit suits on, get set up for the big event, and spend the last little bits of time left on Earth before your life was changed. We were all talking about the exciting experiences we would have on this grand joyage when we were called into the dressing room.

"So, how's the weather prediction for launch?" I asked one of the attendants.

"It's decent, but don't you worry! We won't have Apollo 12 have a repeat for you," he said.

"Well that's comforting, but I'd love a more specific answer," I asked while spitting my gum out.

"It's a clear day and 78 degrees outside," he said.

"Alrighty! Time to make this a historic day!" I exclaimed.

We got suited up for the launch and had a good walk to the tower. We rode the 45 story tall elevator all the way to the top. It was long way down, and none of us talked as the door opened in front of us.

"Matt you're all set! Just climb aboard as the first person, commander," the attendant said. We all got into the rocket and we were excited. I would be piloting us for ascent. Meanwhile at Houston they would be doing the system checks.

"We are go for launch!" said the pad leader."T-30 seconds and counting!"

I stared at the control panel, confounded by all the switches and buttons. The vehicle had four seats, however one person would be left out. The four seats faced the control panel. The control panel controlled 90% of the flight. The docking hatch was just above that. It would be where we would dock to the lander. It was an exciting feeling. This was the first Orion vehicle I would be on! We heard the fuel pumps rumble and everything be prepared.

"We are go for launch! T - 14, 13, 12, 11, 10, 9, 8 , 7, 6, Ignition sequence starts! 3, 2, 1. Ignition!" said the pad leader. The shaking was so intense, I thought I would blow up. "WE HAVE LIFTOFF, LIFTOFF AT 9:45 AM HOUSTON STANDARD TIME AND ORION 9 HAS CLEARED THE TOWER!"

We were off! "Altitude is on the line Houston. We're looking to have a good flight to orbit," I said. This time in the flight was by far the most critical. If just one little glitch occurred, it could blow up the vehicle to tiny bits. Everything looked secure so far, so we went ahead and told Houston about our SRB's and all that was left in them.

"Alright Houston, we're coming up on burn out of the SRB's currently. Standing by for info for jettison," I said.

"We see your SRB's are stable Orion 9. Standby for sequential logic for staging to come up," Houston said. Just then I saw a flashing of a malfunction light on the board.

"Houston, this is Orion 9. We have an engine cut-off, but go on the other 4!" I said. At this point my heart stopped with craze. If we had a flare out of an engine, what were we to do? Well I knew what we'd do. I just didn't want to think of it.

"Houston, what's the story with middle engine 1?" I said.

"Roger that, you still seem to be holding on a good trajectory out into orbit Orion 9. Just keep those engines a firing!" Houston said. "Also, stand by for SRB separation."

"Copy that!" I said. Just then we heard a big bang as the SRB's cut out. I then flipped the staging switch and separated the SRB's. The vehicle was on a good course towards our target.   Our vehicle was working like a charm. Everything looked good, the main engine had flared up again, so we had more thrust than was expected on this flight.

"Houston, we're standing by for main stage shutdown and separation," I said.

"Copy, engine cut off at 9 minutes, 46 seconds," Houston said.

"Roger, the vehicle is working like a charm! We have main engine shut down Houston!"

"Roger Orion 9. Get prepared for rendezvous and docking tomorrow!" Houston said. We had made it through launch. We were in orbit and on our course for the MEM.

"Alright Houston, we're firing up the engine to set ourselves on course for the lander," I reported to Houston.

"Copy, you're looking to have a close approach of 0.1 kilometers away. That's as close as we can get it," said Houston.
We shut off the engine and waited for 1 1/2 hours. We were traveling about 25 m/s in target velocity.

"Slow us down, we're coming in too fast!" I said. We fired up the engine and slowed down to 0.0 m/s. "Alright Reed, put us docked." Reed translated up and down and side to side to get us into position. We lined up and thrusted forward at 0.3 m/s. "Watch it Reed, you're translating up and away."

"Yes I see. Oh great, we have an RCS port stuck on. Matt, we're gonna have to do a spacewalk after this," Reed said. The docking was pretty tense after we found out the RCS port had been stuck on. However, being the experienced pilot Reed was, he had encountered this situation in other flights. So we knew he could get us finished and docked. We heard scraping, and were connected. I went out and repaired the port, and then we did our burn to Mars. It was relatively simple. It took 30 minutes for the Atomic motor to allow us to pick up enough speed to put ourselves on our trajectory. And we finished the burn perfectly! Orion 9 was finally on course for our destination.

Chapter 3

"Alright Houston, we've extended the solar panels and rotated so our antenna gets the best signal and so we charge up the most," I reported as we spun about 98 degrees left.

"Copy, we see you turning around. Make sure not to go out of control on this turn!" CAPCOM said jokingly.

"Hah hah, very funny," I said. He was making a reference to my flight on space shuttle Endeavor. I piloted it and it spun out of control on close approach to the space station.

"Alrighty Orion 9, go ahead and head into the MEM. We want to check and see what system's we have that are online and if the major things are online. If we have something out, we need to abort ASAP," reported CAPCOM.

"Copy that, I'm heading into the lander right now," I said as I opened the hatch to the lander. There was a compression force that shot me into the lander at quite a speed. I hit the side of the lander hard, however thanked the heavens for no punctures in it. If we had a tear, it would enlarge in a split second and would suck me out into the vacuum of space and kill me. Anyway, I turned to see the lander and the vehicle that would make history. The lander comprised of 3 seats. If we had 4 people on the mission, the other person other than the pilot (usually mission specialist) would be in that seat. The lander wasn't much bigger than an average bedroom, and was only as thick as a couple layers of cereal boxes in some places. The control panel faced the seats, and had all the things we need for landing.

"Alright Houston, starting up the MEM soon," I reported.

"Copy, we see it powering up." It usually took about 3 to 4 hours to start up the MEM, so we had to sit still and do it. We were going to float for 5 months anyway, so we might as well use up the time.

3 hours later, we finally got the lander up. All systems were go, computer, map, tanks, everything. We saw some slight scratches in the windows which made sense, because it had been floating up in orbit with space junk coming toward it. At least it wasn't a big crack, or the atmosphere of Mars would tear it open.

"Ok Houston, the lander looks good. All systems are go," I said

"Copy, we see the same here on the ground," they reported.

Our flight was pretty boring from there on out. For the next month we passed out of Earth orbit and sent ourselves going out to pass the orbit of Mars. Once we did that, then we could burn ourselves into orbit, get it's slope perfectly equatorial, and circularize our orbit. Once finished we would prepare for undocking and landing. We would get dressed in our IVA suits and get ready for our big bang. Sean and I would float through the capsule hatch into the Martian lander. We would button up the hatch (and Reed would button up the hatch in the Service module) and then we would undock.

"Alright Houston, we're doing a spin to get better communications with you," Reed reported to Houston as he rotated the craft so that it pointed towards Earth more.

"Copy Reed, we see you rotating around," Houston reported as they talked about the stirring of our oxygen, nuclear, and liquid fuel tanks. Stirring tanks is something that you don't normally do on Earth. Space is different however. In EVA (extra vehicular activity), just moving one foot can change your temperature outside from being 450 degrees to -450 degrees. So, in order to keep the tanks from freezing up the fuel so Houston can get accurate readings (and so we don't spit out ice as fuel), we need to heat it up and stir it around. We do this with a fan and with a heat coil, which allows us to heat up/cool down the fuel and stir it around. We do this regularly so that we don't have the problems listed above.

"Ok Reed, we'd like you to do a stir on your Nuclear tanks, Liquid fuel tanks, and Oxygen tanks. We'd also like you to stir the lander's tanks," said Houston.

"Copy, I'm doing the Service module (The main area, launch area and control panel are in the command module, which sits attached to the service module) and Sean's doing the MEM," Reed reported.

"Ok Houston, I think we're gonna get some sleep here and eat," I said.

"Copy Matt, good night commander," CAPCOM jokingly remarked. And so it was that we all ate and had a good time talking. I went to sleep and left Sean in charge. Eventually, Sean went to sleep and woke me up and told me Reed was in charge. So I went back to sleep till Reed woke me up and told me he needed some sleep bad. I don't know why all of us didn't just sleep all at once. Maybe it was in case we had a problem we wouldn't be blind to it. In fact we DID have a problem.

I heard the buzzing of an alarm and turned in horror, knowing it would be a LONG flight back if something happened. "Uh Houston, we've got a slight electrical problem," I moaned at them.

"What's going on Matt?" Houston said. My crew was up and ready for action by this point.

"It seems we have an overload of power and gimbal lock on the engine for our service module(called Lil' Dipper)!," I stated.

"An electrical overload?!?!?!?! What systems are offline?" Houston yelled back.

"We've had batt's B and C overload. I don't know what happened, but it seems like the solar panels are overloading the power," I calmly reported.

"Copy, we're thinking of something down here," they said.

"One idea Houston, we need all the things turned on right now. If we can, they may draw enough AMPS to set the balance straight," I said.

"Copy, we're thinking of something else. Retract the solar panels on the lander. If you do that, you may be able to get it equal," they said.

"Copy, we're doing that now." We turned off the panels in the lander and all was well. Something that worried us however was the shock that would be produced when we undocked with it. If we had too much power, the vehicle would short out upon undock.

"Houston we're going to draw some AMPS out. We're going to turn on all systems and suck up all unneeded AMPS," I said.

"Copy that, sounds like a plan to me," they said. Our flight was smooth after that. Every day it was the same thing. Wake up, brush teeth (yeah, couldn't go that long without brushing), make a course correction of less than 10 m/s of DELTA V, change position of craft every so often, talk to Houston, read a book, play some music, eat, sleep. That was how our days went for the first month. It was a good save of the craft that day. Everything looked good!

Chapter 4

"Well boys, it's been 1 month since launch," I said. "We've been doing something right apparently. All other flights got sick around this time."

"That's true of course," Reed said as he rehydrated a bag of spinach. "Maybe it's because we're the first Orion flight without a super long flight problem list." That was true. Orion 7 (a flyby of Mars) had an Apollo 12 pulled on them. They got struck by lightning on liftoff, but thankfully were able to pursue the mission fully. Orion 8 obviously blew up at lift off (always a traumatizing experience). Orion 6 was the first time a human saw the surface of Mars, as they orbited the Red Planet 10 times. Orion 5 landed on the Moon, as well as Orion 4 and Orion 3. Orion 1 was an orbit flight around the Moon, and Orion 0 was the first Orion flight to fly (unmanned). The government loved the idea of Orion, so NASA (and us as well) thought we could get up to Orion 56.

Our flight was pretty simple after the escape from Earth. I constantly traveled through the tunnel into the lander and stared at the control switches. Everything seemed relaxed and cozy. No serious mishaps had occurred since the electric problem. The most serious problem on any Orion flight (that they lived from) was the LEM (Lunar Excursion Module) getting hit by a meteor. It crumpled in on itself. They buttoned up the hatch and detached the LEM ASAP. All was well after that. That was Orion 2. It was meant to be what Apollo 10 was essentially. It would fly close to Lunar soil, then detach as if an abort occurred. The situation thankfully was fixed, however Orion 2 never went to the Moon. All was well on Orion 9 though. The flag was still inside the lander, so we were good on that. The lander was ship-shape as well. Our flight was going smoothly. The solar panels were slightly broken on the ends, so we went ahead and detached the ends. The service module engine looked good as well. We did frequent checks on the vehicle. We had to make sure the heat shield on both ships wasn't damaged. If they were, this would be a short flyby of Mars. The landers heat shield wasn't as important however, so we were fine if a dent or two was there.

All of our communication systems were online as well. We were floating in the vacuum of space with everything working fine (thank the heavens).

"Houston, we're gonna go ahead with another stir on those tanks. Everything looks good and is working beautifully," I said.

"Copy, we see the same here. We don't know what else we can do with the rest of this time right now!" they said.

We were getting very close to Mars by now. The flight there had been a sturdy one. All things were going well. The scenery was getting slightly old though.

"Houston, we're about to pass into Mars' sphere of influence, correct?" I said. A sphere of influence is when you enter into a planet's gravitational pull. When this happens, we will slow down our interplanetary speed and go ahead and be in an orbit around Mars. This manned mission would be going within about 50 kilometers from Mount Olympus, the highest mountain in the Solar System.

"Copy that, we need to do a burn for you to get a closer Apoapsis to Mars," they said. An apoapsis was the highest point in an orbit. A periapsis is the lowest point. When both the Apoapsis and Periapsis are at 0, the vehicle has landed. We did our burn and lower our encounter Apoapsis to 75 kilometers above the Martian surface. We waited to fly into the sphere of influence of Mars, and when we did, our NAVBALL and instruments went bonkers (because of course they were calibrated to be in the Sun's sphere of influence). So we calibrated the instruments to be use to Mars. We then reported to Houston.

"Houston, we've passed into Mars' sphere of influence just now," Sean said.

"Copy, we see the same," they responded.

"We're going to schedule our burn at Apoapsis to burn into orbit," Reed responded. By this time I had gone back to sleep. I woke up about 2 hours later so we could perform the burn. The burn was simple, and the service module engine made quick work of it (2 minutes in fact). We were in orbit around the lost planet Mars. A planet which has it's history covered in red myths. Some believe (including me) that life existed on this barren planet. If that was the case, then we would possibly be able to make a self sustaining base here, with farms inside rockets. The atmosphere is Co_2 like, however the rover Curiosity has seen peebles. What do peebles mean? Flowing water of course! Can flowing water exist in an atmosphere of Co_2? No, only an atmosphere of Oxygen. This means that Mars had the capability to sustain life once upon a time. Also, although we haven't yet, if we drill down at the polar caps on Mars, we may see an enormous amount of icicles and/or ice.

"Houston, the sight here is amazing. All the materials are looking good," I said.

"Copy, we see the same Orion 9. Be prepared, we may deorbit right now ha ha ha," they said.

"Ha ha, very funny. So in all seriousness, when are we going to detach and go to the surface?" I asked.

"Probably in a week or so. We still need to get all accurate readings from the rocket and then we can go ahead and go you guys a going for landing!" they said.

"Copy," I responded with a grin. The next couple of days were pretty good. We just made our passes around Mars with the greatest of ease.

"We're within 75 kilometers from making history guys," I said. They both chuckled and went off to bed. I would've joined them, but I decided to take in the sights from here for one last time before my life would eternally be changed. It was crazy to think that we could actually land this thing. We would be the first men to not only step on Mars, but to stand on a planet that is not our own.

"Alright Houston, are we go for our separation and landing?" I asked.

"Copy, go for your dressing and separation in 2 hours," they responded. "How was your day so far?"

"Fine, thanks Houston," I said. We all got dressed in our IVA suits (the one's we launched with) and Sean and I moved into the MEM. We started it up, got the solar panels extended, and said our farewells to Reed.

"You guys be careful, I won't be able to abort this flight for you guys," Reed said.

"Yes, we know perfectly well that once we detach and fire up the engine, there's no going back till we abort or land," Sean said.

"Anyway, god speed Sean and Matthew," Reed said.

"You as well Reed," Sean and I responded. We got ourselves into the lander and buttoned up the hatch. We assumed Reed did the same thing cause we heard a clang on his end. We started up all the systems in the lander and got the panels primed, however they weren't taking power in yet. Reed was turning on MEM communications in the command module, and we talked to him.

"Alright Reed, you ready to detach?" Sean said.

"Copy, I'm go for detachment on my end," Reed said.

"Well go on then! Let's get this mission done!" Sean said. As he said it, we heard a big bang and were thrown upward. Unlike the LEM on the glory days of the Apollo missions, our lander had seats. We went ahead and used our RCS system to thrust away from the service module. Our lander is called Aquila, so we got referred to as that.

"Aquila, this is Houston. Do you read?" they asked.

"Copy, yes we do Houston," I said.

"You know what to do right now, correct?" they wondered.

"Copy, yes we do. We're going to Apoapsis on the dark side and burning downward into a 15 kilometer entry to Mars," Sean responded. We were to talk to Lil' Dipper today. However, we decided not to yet. Our heatshield looked good, as well as all the other systems. We fired up the engine and made a long burn. We put ourselves on course for entry. We were going to make history.

Chapter 5

We had flipped around to the dayside of Mars. We were in the atmosphere, however we were not on a course for landing. Our flight looked good! We were entering and on course for our target. We started to enter the atmosphere and put our heatshield up. Everything looked good, but it was getting really hot. Our main engine was starting to get a little hot, but everything was good. We lost our communications with Houston. However, Sean and I looked good on reentry!

Our entry angle was a little screwed up, however everything looked normal. We jettisoned the heat shield, fired up the engine, and slowed down for a landing. Our retro rockets were primed and all was good, until one parachute caught on fire.

"Houston, we've had a parachute catch fire!" I yelled.

"Copy, we have 4 parachutes on the lander. Just cut the one on fire and the opposite one," they said.

"Roger that, we're going to burn a little more for landing however," I said. The parachute had burned a little hole in the lander as well, which caused us to freak out. As I stated before, a single hole could expand to a huge tear and suck us out into the vacuum of space.

"Aquila's damaged Houston. We've got a hole in the side. I don't know if it's capable of becoming bigger, but it doesn't look healthy," I told them.

"Copy, we'll do so repair work once we land," they said. We fired up the engine and popped the parachutes. We were committed to land now.

"Houston, we've got the parachutes out, we're firing up the main engine to slow us to less than 3 meters per second for landing," I said. If we weren't slow enough, the lander would blow upon impact.

"Copy that! We see you coming close to the surface," they radioed back.

"Houston, we're 1 kilometer from expected surface!" I said. We were 1 kilometer from making history. The retro rockets were charging to fire.

"Aquila, go for shutdown of main engines and firing of retro rockets in 3 seconds!" Houston said.

"Copy that, engine shutdown, retro rockets primed for ignition in 3...2...1...Ignition!" I said.

"Aquila, you're currently 40 meters from the surface, and going 2 meters per second vertically," Houston stated. "Aquila, at this point in time, you should be on the surface. Do you copy?"

No response. Everyone in Mission Control was extremely nervous and feared the worst.

"Aquila, do you copy?" they asked. "Aquila, this is Houston, do you copy?" "Aquila....are you there?"

"Calm your horse's Houston! We're down on the surface and see Olympus Mons off in the distance! Man, our hearts nearly were stopped there!" Sean reported. Cheers were heard in Mission Control, and millions of people watched in awe as we talked to Houston back and forth.

"Ok Houston, we've started repairs on Aquila. It was a pretty big hole in the side," I said

"Copy, we hear you. Well, do you have duct tape?" they asked jokingly.

"Hah hah, yes we do," I said. "Houston, can we go for an early EVA?" I asked.

"Sure thing Aquila, everything looks fine. We're going to ask around here and see if we are go for staying on the surface," they said

"Thats fine with me, just make it fast," I said. In Mission Control they would be asking around for go/no go for staying. This happened on Apollo as well, so we had to keep with tradition I guess.

"Aquila, this is Houston. You are go for staying on Mars, as well as the 3 hour before scheduled EVA," they said.

"Copy that Houston, we're going to check with Lil' Dipper, as we see it flying overhead," I said. I then radioed to Lil' Dipper. "Lil Dipper, is everything go on your end?" I asked.

"Copy that Aquila! You've got a smiling face up here in the Command Module!" Reed exclaimed happily.

"Hah, and we have two down here on the surface!" I exclaimed. We had made history as the first men on the surface of Mars.

"Houston, the dust seems very different than anything I've seen. We're scooping up a sample and getting it inside the lander," Sean said.

"Copy that, we're getting some pictures from you guys of the surface of Mars and of the soil," they said. We examined the soil with the greatest care, however realized that it hurt to touch with the bare hand.

"Houston, we've tested how the soil feels and how our body reacts to it with bare skin," I said. "Our studies have shown that it seems to be a red type of ice, however it seems closer to Dry ice in effects."

"Copy, so it burns to touch with the bare hand?" they asked.

"That's affirmative Houston," I replied. "It has a slight burning sensation upon feeling." Our next 3 hours were studying the soil, and once we approached 2 hours till EVA, we started getting dressed in our EVA suits. Space suits take almost 2 hours to put on, even if they are brand new. Our EVA suits were new because they were designed to take more force from space junk.

Chapter 6

We were dressed in our space suits and were primed for our EVA.

"Houston, we've done a lander depress of the cabin," I said. A lander cabin depress sucked all the air into the outer walls and kept it there until we "undepressed" it. If we didn't do that, the air would force us out of the hatch and destroy the lander, so it was pretty important.

"Copy, you're go to open the hatch," they said excitingly.

"Roger, the hatch is opening slooowwwwlllyyyyyy," Sean said. I would be heading to the surface first, along with the flag. I grabbed the flag and put it on the side. I imagined the whole Earth (couldn't say world, or I would be suggesting life on Mars without scientific facts) staring at us while we were climbing down the ladder.

"Houston, I'm on the first rung of the ladder now, Sean started up the camera just now as well," I said.

"Copy that, we're starting to get images and video from you Matt," they said.

"Houston, I'm on the last rung of the ladder now. The dust seems similar to the Moon's, however it's kind of a redish brown color. I know I'm killing you guys with the suspense, so I'm going to step of the MEM now," I said. I took one last look at the MEM (my last moment before being utterly famous and have too much attention!) and looked at the ground. I stared at it for a good ten seconds, and got prepared for this jump. I then looked at the pad, and lept from the ladder.

Silence was heard in Mission Control. Then, crackling cut in as the radio was turned on in my suit.

"Houston, this place is a place of tranquility, a place of safety, and a place to admire everything we know about our lives," I said. "I'd like everyone who has helped us get here, either with encouragement, commanding us, or anything in between, to take a moment and contemplate the events of the past 3 hours. I'm no God, but I am one of you. I am a man of flesh and blood, and am no better than any of you. Keep this in mind, as you see me place the first step of mankind on Mars," I said.

"Houston, as Neil Armstrong once said,"That's one small step for man, one giant leap for mankind," I would like to say this again. We have lept over hurdles and barriers to reach the impossible," I said. I took a look at the scenery and was taken aback by everything on this once called barren planet. Suddenly, I gasped. I thought I was going to have a heart attack!

"HOUSTON, WHERE IN THE HECK ARE YOU?" I yelled. "There's life on Mars! I see ants not unlike ours at Earth here!"

Everyone in Mission control was taken aback by my statement. Life really existed on Mars. Thriving life no less.

"Copy, you know what we will ask you correct?" they asked.

"Copy, I'm taking my space suit off right now," I said. I slowly took my helmet off, just to make sure. As the helmet repressed and the air hissed, It seemed like nothing that we thought would happen in normal space happened. I took off my entire EVA suit and dressed in my normal flight clothes. I walked around like normal! Everything was yankee doodle as it were. Sean took off his EVA suit as well, and got dressed in his normal flight clothes. Everything seemed fine! I planted the flag on the planet and all was well. Our next 6 days on the red planet were as smooth as could be. We saw Olympus Mons erupt a couple times (a nice treat) and saw more ants and the remnants of what seemed like metal (like the stuff used in railroads). After our couple of days, it was time to head home. Our flight home would take two passes around the Sun, so it took much longer. Our flight home would take close to 8 months, as we were going much faster this time, because we were in a higher orbit. We would wait for Earth to be in the right position for us to burn out. Once we burned out, we just had to wait for Earth to match up with us and wait for quite a few months. It would actually be pretty boring. The same old stuff every single day. Boringness, it's a cruel mistress.

"Alright Houston, we see Lil' Dipper in the correct position. We're going to burn up to it soon," I said.

"Copy that, I see you guys in the right position," Reed said.

"Ok Houston, engine ignition in 5...4...3...2...1...IGNITION," I yelled. I fired the engine up and Sean started piloting us up to get into our rendezvous position. Our flight had been a rather successful one so far. Only one real problem had occurred with Aquila.

"Houston, we're holding a good heading and are picking up speed through the atmosphere. Also, we're standing by for jettison of the lower stage and ignition of part two," Sean said.

"Copy that, everything seems good on our end as well. However, Sean, you may want to slow down slightly. Your speed is getting dangerously fast," they said. Unfortunately, Sean didn't hear that last part, so we were heading up through the atmosphere at 400 meters per second. The problem was is that this was about the speed that your speed (in the form of energy) gets dissipated off and outwards, causing flames. This is also known as Entry-Interface, or re-entry.

I heard the beeping of an alarm, and I could tell that Sean heard it as well. I looked at the panel to see if we could solve what was going on, however I couldn't see the panel well because it was vibrating so much. I looked out and saw flames licking at the side of the MEM, and I was stricken with fear. I saw the oxygen level decreasing rapidly. Nothing was working like it was intended to.

I yelled into the microphone,"Houston, we've got an extreme problem with Aquila here!" No response. I yelled again. Once again, no response. By this point I was getting extremely nervous. It seemed like our communications system was offline. We couldn't track anything, our power was next to dead, our SAS and RCS systems were dead, cause SAS needed power and RCS required Mono propellent, which happened to be drained. Aquila was bleeding to death. I yelled to Sean to slow down Aquila. If we didn't do so, the pressure that was being given to Aquila already combined with the force of the engine would tear it apart and kill us. It would be rather gruesome if we died that way as well.

"I hear ya Matt. I've cut the engine power and we're coasting up on force alone," Sean said. Aquila was still burning, but thankfully we were still in the gravity well of Mars, so we could fire the emergency entry sprayers. Thankfully it put out the fire, but Aquila was badly charred. One little tap would destroy the craft completely. We had to be careful where we were for the next bit, because rendezvous couldn't be done by Aquila in this condition. Lil' Dipper had to do it.

"Sean, fire up the engine slooowwwlllyyyy and burn us into orbit," I said. I didn't want to risk a hole forming and break up. We fired up the engine at 5% thrust for 3 minutes and got ourselves into orbit.

Chapter 7

"Alright Matt, let's get strapped in and start orbiting and waiting for Lil' Dipper," Sean said.
We couldn't communicate with Lil' Dipper and Houston at ALL.

"Alright Sean, you're lucky I haven't killed you yet. This could have costed our lives," I said

"Well I'm *sorry* that I messed with your beautiful plans and got excited to be reunited with Reed, Matt!" Sean snapped.

"Don't you talk to me like that Sean! I'm the experienced one here and know what I'm doing! That's why I'm commander you idiot!" I yelled back.

"Well why are we appointed to be on the same flight then, ay?!?" Sean yelled toward me.

"I don't know, but I know I'm not going to be on the same flight as you again!" I yelled.

"Fine then! Since you don't want me, I'll crash myself into Mars!" Sean said as he got up to go toward the hatch.

"SEAN, NO! GET BACK HERE!" I screamed. It was too late though. Sean had opened the hatch and had left me hanging by a chair while the air whooshed out and threw him into deep space. I snatched a rope that was hanging from the wall near by and tied myself onto the chair. I closed the hatch, decompressed the chamber, and reopened the door. I lept out into deep space, and looked for Sean.

From experience and study, his suit should have lost communications which made it harder for me. However, I started up my computer (I had the sense to put on my EVA suit instead of IVA suit) and looked for the last split off of a second ship. The records were junk, just showing the last ones as when we jettisoned from Lil' Dipper. I had to find him manually, and had to do it in less than 5 minutes. That's how much oxygen that the IVA suit had (not attached to the tank for liftoff).

I saw a little flash of white, which was hard to spot because of all the stars. I thrusted down with my pack fuel and reached Sean in 4 minutes. He was unconscious, and appeared to be in shock. I thrusted up to Aquila again and got in with moments to spare. I shut the hatch, recompressed the room, and got oxygen going into it again. I then took off his helmet and IVA suit, and proceeded to do CPR in a similar fashion to how you do so on Earth.

CPR is difficult to do in space, because there's no gravity that you can use to press down on someone. So I had to grab something that would hold us both in place. I used one of the seats that we had crushed into a bed. I strapped him in, strapped myself to the wall, and started slamming my knee into his chest. This was the only real way we could simulate what was done with CPR on Earth. I pressed for 10 times, then breathed in twice. I repeated this several times until he started to stir and wake up.

Sean coughed, and then slowly got up. Oxygen was going through his limp body again. I breathed a sigh of relief and started to get out of my EVA suit. After Sean was recovered and I was back in my Orion

9 commander "utility" shirt, we went ahead and started talking again.

"I'm sorry, I was a bit immature there," Sean scratched out. He could barely talk, and it was quite an effort. I could tell he wasn't going to be able to do much for the rest of the mission back to Earth.

"Me as well, I should have handled the situation a little better," I said. "You look tired, go and get some rest. I'll be at the control panel," I said.

"Ok, I'll get strapped into our makeshift bed," he said. He was right about the "makeshift" bed part. It was essentially the other seat that was not going to be filled torn into a seat. It was a decent bed, rather comfy as well. Sean got strapped in, and I said goodnight. I then turned toward the control panel and watched the time to close in tick down. Everything was semi-back to normal. Again, no communications were online on the MEM. We're hoping against all odds that nothing happened to Lil' Dipper.

Everything seemed decent. The time to target was not increasing which was good. However, something even WORSE than our current situation happened. As Sean was sleeping and I was staring at the panel, everything zapped out in the cockpit, and I flung upwards with a start. Nothing was online, except for the Oxygen system and Co_2 filters.

"Sean, wake up now!" I said.

"Huh? What is it now Matt?" Sean said. I didn't even have to say anything. He looked forward, and turned pale.

"Well, if I wasn't disciplined, I would be cursing my brains out right now," Sean said.

"Me as well. Let's see if Lil' Dipper can get in. I still have 97% on my phone, let's see if we can time it for 1 hour," I said.

"Copy, all we can do is wait," Sean said. For the next hour, we tried to keep warm, get some sleep, and try and make the best out of the rest of this mission.

Meanwhile in the Command Module, Reed was speeding up to account for the venting that the door did when Sean opened it.
"Houston, this is Lil' Dipper, do you copy?" Reed said.

"Roger Reed, we hear you loud and clear," Houston said.

"Ok good, I couldn't get any signal from Aquila, so I was wondering if the communications system was offline," Reed said.

"That's affirm Reed, we've had the same problem," said Houston.

"Copy, do you think systems are down?" Reed asked. He got no response. "Well, gotta do this without any reference points." He fired up the engine, and got within 10 meters of Aquila.

"Aquila, this is Lil' Dipper. Do you copy?" Reed asked. No response. He decided to get docked and then go ahead and talk to them. He thrusted up and around into position, and then thrusted forward.

"Approaching at 0.3 meters per second," Reed said out loud. "Thrusting forward."

"Houston, we have capture!" Reed said. Sean and I could hear it as well. We were docked!

"Reed! Decompress the command module! I'll do the same here!" I yelled. I decompressed Aquila and Reed decompressed Lil' Dipper. I opened Aquila's hatch, and Reed opened Lil' Dipper's. Sean floated through and was glad to be reunited with Reed. I radioed to Houston.

"Houston, this is Matt reporting from Lil' Dipper. We are docked and are going to head home!" I said. Houston was extremely happy, and gave us our burn data to escape Mars. We performed it and barely were able to have enough fuel to put us on a two pass around the Sun (to get to Earth) orbit. Our flight was relatively simple, however Sean and I were paranoid about the MEM getting a hole and killing all 3 of us. We kept the hatch close to the door. Sean and I did a space walk and checked out the MEM and what was going on with it.

The results were shocking. The whole front of the craft was ashed up, and everything was charred and destroyed. The control panel was mainly hanging from a couple of wires and there were many spots that were destroyed. Everything on Aquila was broken. It was clear we couldn't start the craft up again. Unless we somehow were able to repair this thing, it was a gonner. Sean and I went back in and he radioed to Houston.

"Houston, are we allowed to just dump Aquila? It appears to be unfixable," Sean said.

"We'd prefer you didn't, just in case something happened to Lil' Dipper," they responded.

"Ok, I don't know what we could do with a pile of junk like this if Lil' Dipper broke," Sean said.

"Copy, just have Aquila hang on there, it's not that big of a deal," Houston said. So we left it on and stared at Mars for the next five hours, watching it float off into the distance. Our flight had been successful, with a few minor bumps like all flights. We flew out of the sphere of influence of Mars and went into orbit around the sun. We passed for our first time around the Sun, and thats when Lil' Dipper was getting annoyed. It was ticked off at us, and didn't control with us like we wanted it to. By this point, Aquila could barely control it's life support systems, however it was online because of Lil' Dipper controlling it. We started our second pass (it was one year since liftoff) and then something fatal happened.

Chapter 8

"Lil' Dipper, we'd like you to perform a heating of the tanks yet again," Houston said for the what seemed like millionth time.

"Copy, I'm doing that now," Sean said. He reached for the stirring/heating switches and flipped all three. As he did so, the craft jiggled and wobbled. Suddenly, Lil' Dipper exploded with the sound of alarms. All systems were not functioning. Red lights were flashing all over the place, and everything was failing. We saw electric charge go from 100% to 5% in 5 seconds on batteries A, B, and D. Three of the five batteries failed. Without them, we wouldn't have enough power to get back. All of the nuclear tanks were drained, the liquid fuel was creeping down, and Lil' Dipper's lights were flashing on and off, which meant we were close to having a shortage.

"Houston, we've had a pretty big bang here on Lil' Dipper. I don't know what's gone wrong, however I do know that we've got a critical failure here!" Reed said.

"This is Houston, say again please?" they said.

"Houston, we've had a major malfunction here!" I said shakily, knowing the real problem they couldn't see was down here. I saw something that seemed like a little flare going off. It was too bright and too strong to be RCS. Suddenly, I just realized what was going on. The failure's weren't an engineering fail. They were something worse. Something that nearly costed the lives of the Apollo 13 astronauts.

"I'm going to reconfigure the RCS guys!" Sean said.

"I've rolled the craft back into a semi controlled state," Reed stated calmly. He was barely keeping himself under control though. I could hear the quivering in his voice, as well as mine.

"Crap, flight?" the surgeon in Mission Control said to the flight controler.

"Go surgeon," he responded

"Their heart rates have zoomed to 198 beats per minute. Thats unhealthy, even after exercise," he said. He was right of course. My heart was beating to death, as well as my crew. If we didn't watch ourselves, we could have a heart attack and have a ballistic course down to Earth. Or hopefully down to Earth. With the little flame outside, I had no idea whether we were holding our entrance into the SOI of

Earth (SOI of course being sphere of influence). I decided to tell them about the flare.

"Houston!" I said. "We're creating some junk out here! Something is pushing us off course, but it's not RCS. It appears to be a gas and fuel with electricity to set it. It appears to be….. oh…..no…….Houston…" I was afraid to respond.

"What is it Matthew?" Houston said as I stared at the little flare for a while.

"Houston, it's oxygen and our liquid fuel. The nuclear fuel and bombs already went off, which may mean the service module is infected with radiation. Either way, I'm nervous to go down there. Our Oxygen tanks 1 through 6 are reading zero. Our others, 7 through 9 are reading 50%," I said. We didn't get a response from Houston for a good 3 minutes. "Do you copy Houston?"

"Roger, we copy you venting oxygen and fuel. Go and try and refuel the remaining tanks that don't read 0," they said.

"Copy, we're doing that now. At the same time, we're going to power up Aquila somehow and power DOWN Lil' Dipper," I said. "Can you give us a time estimate of how much we have on life support here Houston?"

"Do you really want to know Matt?" they said.

"Not really, but what are the numbers," I said.

"We're looking at 20 to 30 minutes of life support in Lil' Dipper," they said.

"Well, crap. It takes us 2 and a half hours for just the essentials to be turning on with Aquila," I said.

"GUYS GET A FREAKING MOVE ON!" I yelled. "WE'VE GOT LESS THAN 30 MINUTES TO POWER ON AQUILA!"

"Roger Matt! I'm in Aquila starting it up and getting this charred mess to start up and have them panels extend," Sean said. Then I facepalmed. I remembered, if we had solar panels, we could use them to charge up the craft.

"Houston, we're leaving Lil' Dipper on. However, we've decided to close the valves on the tanks that are leaking," I flipped the switch to shut off the valves.

"You realize that that is not able to be undone. Correct?" they asked.

"Yes, I was a commander of the ISS and Space Shuttle Atlantis twice. I flew all of the space shuttles except for Challenger, so I know what I'm doing," I said. Challenger was a space shuttle that (at the time) was the most reliable rocket of them all.

We started Aquila and got the panels out ASAP. We also got the engine primed on Aquila. We weren't lighting Lil' Dipper's engine on my watch. I assume we ruptured the hull. That made me nervous, cause we could have just lost our lives. I went ahead and told Houston that the batteries on Lil' Dipper were holding low, however the fact they were holding at all suggested we were doing something right, right?

"Houston, we've got the panels out currently on Aquila. We've got the electricity coursing through everything now, and we're looking good," I said. Suddenly I heard a clash of alarm noises. It sounded like someone exploded the rocket. This wasn't good. We certainly had left our trajectory to enter the SOI of Earth. Our flight wasn't looking good.

I flew into Lil' Dipper to see what was going on. I thought I was going to have a heart attack. All of the alarms lost control. We were spinning out into deep space and were starting to get very sick.

"Houston, we've had everything in the world drop out in Lil' Dipper! The blast is opening a huge hole in the Service Module!" I said.

"Copy, we see the same here," they said.

"We see an emergency shut down light on Houston! Oh gosh, there go 5 more! CRAP, Houston, we've got a full emergency shutdown light on!" Reed yelled.

"Copy, transfer all gimbal angles and computer info to Aquila ASAP. We need to get it started up or you guys are going to be lost in space for quite some time," they said.

Chapter 9

"Houston, we're getting our systems up here. We've transferred all info over to Aquila's computer," I said.

"Copy, 30 seconds to emergency shutdown." Sean was hastily scribbling down numbers to enter into Aquila's computer. Reed was powering off things once it was finished. He was intending to conserve the power for all of the other systems. I was yelling packing up all the stuff that we would need for the trip home. I shut down the fuel tanks and oxygen tanks on Lil' Dipper and transferred it to Aquila. It would have to do for now, until we were settled in with our situation. Right now, we weren't at all.

"Lil' Dipper, that's 15 seconds till shutdown....5...4...3...2...1...We have loss of contact with you Li.." Houston's chat with us was cut off with the shutdown of Lil' Dipper. We were off to Aquila again.

A faint crackle was heard. Then we heard Houston say,"Welcome back Aquila." In the meantime, Reed was sitting in the Command Module.

"Houston, do we know for sure that we can power it up again?" Reed asked.

"Copy Reed, we'll just take that one at a time," Houston said.

"Copy that Houston. This is Lil' Dipper, signing off," Reed responded glumly. He floated through the cabin into Aquila, and looked at us while Sean and I were talking and trying to control Aquila with a dead rhino on our back. After a while (once we were under control again) Sean turned to Reed.

"Sorry I broke your ship," Sean said.

"It's fine, but you could've just broken our chance of living as well," Reed said.

"Alright Sean. It's fine that we've lost one ship, however, if we lose Aquila, consider ourselves dead. We need at least one ship to power up the other ship. What we need to do is burn ourselves back into a trajectory to Earth's SOI before the morrow. I'm not stirring those tanks again," I said.

I was burning the engine and sending us on a re-entry course for Earth. Push comes to shove, we could use Aquila for re-entry, but it could lead to a disaster. If the emergency heat shield doesn't last (because it was meant for Mars) we would have a fiery death. Also, the parachutes on Aquila already have been deployed. They'd be weak if we re-deployed them.

"Houston, our flight path is looking good. I'm going to try and get some sleep here. I'm checking out the service module as well, to see if there isn't any radiation in it," I said.

"Copy, what do you see in the service module?" they asked.

No response. Houston lost contact with Matt.

"Reed or Sean, can you check the service module and see what the devil is going on with Matt?" Houston asked.

"Sure, I'll do it Houston," Reed said.

Chapter 10

<u>Side note: The story is now told from Reed's perspective. You will find out why later on.</u>

I floated through Lil' Dipper, which was very dark right now and entered the Service module. The radiation alarms were online, however it didn't seem as if it was in there, cause only 1 alarm was on. I went around the hallways and into the Commander's bedroom. This is where we slept for most of the flight. I entered and was shocked to see Matthew inside of his room, floating and sighing a deep sigh.

"Matt, are you ok?" I asked.

"No, I'm not ok at all," he said. I was nervous. What happened to our commander?

"What happened?" I asked he moaned and told me.

"I entered my room like I always do, and found that my computer decided to "shock" me. Yes, pun totally intended" Matt said.

"Will you die?" I asked

"Yes, I will pass into the next life on this flight. Do not feel that you must take my body to Earth again," Matt said.

"I will Matt, even if it costs me my life," I said.

"Thank you Reed. Take my place now," he handed me his commander pins and shirt.

"Copy. Rest in Peace, Matthew Psy," I said. I left his body in his room, with him in his sleeping bag tied to the wall. I floated back into Aquila with the commander shirt on.

"Oh, look who it is! It's Reed! With a...wait what?" Sean said.

"I've been appointed commander of Orion 9 for the duration of the flight," I said.
"At least by Matt I have."

"What happened to Matt?" Sean said, worried.

"May he rest in peace. He has passed into the next life," I said. Sean gasped. He went straight to his room to see for himself.

"Aquila, do you know what happened to Matt?" they asked.

"He has passed into the next life," I said. Houston said nothing. We've never lost an astronaut in space, especially in that kind of way.

I went around the craft to look for something to do, but couldn't find anything. I started talking to Sean who was glum in the service module.

"Sean, how ya doing?" I asked

"Tired, and not really entertained," Sean said.

"Well, we can go on a space walk and see what's going on with the craft," I said.

"Sure, let's ask Houston," Sean said. We asked, and they said yes. We got our space suits on and decompressed the craft. We stared at each other while I opened the hatch. I lept from Aquila and out into space. I turned to look at the service module and was horrified with what I saw. An entire panel was blown off, and 90% of the service module was torn up. Everything looked as if a rabid monster chomped and bit at the craft. Nothing seemed hopeful. Without the service module, we'd lose the craft completely.

"Oh geez. Houston, we are getting our first look at the service module right now," Sean said. "It appears as if something crashed into the craft, making us lose control and systems."

"I suggest something darker Houston," I said. "It appears that we've had an EXPLOSION in the craft." Houston was dead silent. No breaths were drawn in Mission Control. They just got a new idea on everything.

"Copy that Aquila, we register your blow out," they responded. We decided to shut out the space walk after piecing together what we could on the craft so far. We put on some metal plates on there and went into Aquila once again.

"Houston, we're going to close out Aquila and sleep for the night in the service module. Let's hope that radiation doesn't kill us," I said while yawning. I went to bed and started to talk to Sean. Matt's body was inside his room, and it was rather sad.

"Sean, how are ya buddy?" I asked

"Depressed. You'll be a good commander for the rest though," Sean said. We went to bed and started to do our sign off stuff in our computer. I signed off and started browsing. It was pretty cool, and Orion is much different than Apollo. We can talk to our families in this thing. The thing of which I talk about is of course Orion. Our flight is going decent so far, thank the heavens that we're going to be safe here.

The next morning, I got up and signed in. I then knocked on Sean's door and told him to wake up. I started to log up the data for Lil' Dipper so far, and things were looking good. We made it for about 3 days by this point, because after the burn to put ourselves out into our trajectory to go inside Earth's SOI we went ahead and waited. THEN Matt went and checked with the service module. That's when he died. We had lost our main commander, and it was sad obviously. I attempted to start up Lil' Dipper, but it refused to go online. We needed to get it online ASAP. Sean was now the Command Module pilot, and I had to show him how to control everything. We had to make sure that we could survive this flight. We went ahead and started up the rest of our systems on Aquila. We have SOME systems up on Lil' Dipper thankfully.

"Houston, we've got some systems up on Lil' Dipper, a major accomplishment over what we had!" I said.

"We see some systems up on Lil' Dipper here as well," Houston said. "Also, it's agreed. It's a major accomplishment."

"Copy that! We see our computer just peaking over the edge with startup!" I said.

"Roger, we see the same," Houston said. I hollered down to Sean to get his butt up here and see what I was doing.

"Wow! We've finally got my ship now (irony, it's the spice of life), so we can get everything WORKIN'!" Sean said.

"Copy that! Houston, we've got Lil' Dipper mostly online! We still have the tanks shut down however," I said.

"Copy we see the same here Aquila!" Houston said.

"Copy Houston, this is Lil' Dipper, resigning on!" Sean said. It was amazing to me to see our flight turn around so quickly, and then turn BACK around!"

"Roger, we agree with you," Houston said. It was a pretty easy day from that point on. we just did routine activities, including repairing the side of the service module again. It was such an easy day that I started to get bored. I started playing "catch" with Sean by throwing a tennis ball back and forth between Lil' Dipper and Aquila. We did that for about an hour before I just went to my computer and started chatting with my family. Suddenly, I hear a weird voice behind my head.

"You shouldn't have done that….," the voice said.
"Who goes there?" I yelled.

"I will give you one more chance….to go back in time. Will you take it, or will you accept your fate..," the voice said.

"If it's you Matt, then I will do it for you. If it's to experience torture again, then no," I said.

"Very well, feel free to lose all that you have from your grasp….," the mysterious voice said. Although I thought it was someone pulling my leg with this whole voice thing, I was still nervous about it.

Chapter 11

After about 6 months of floating around in deep space, we were prepared to enter the Earth's SOI. It had been a painful time, as many systems broke on board Lil' Dipper and Aquila. At one point, the communications system was down in BOTH Aquila and Lil' Dipper, leaving us stranded for nearly a month with no communication to Houston.

We had done it. We had passed through our flight and completed our mission with only one casualty so far.

"Aquila, stand by for entrance into Earth's SOI. 3...2...1...Welcome back to the gravity field of home!" Houston said.

"Copy that Houston, thanks for helping us through our treacherous flight. We're going to get primed to go for our re-entry in 2 days," I said. We performed a burn to put our apoapsis down to 20 kilometers from the surface of Earth. This would bring us down fast and steep through the atmosphere, so we had to be careful upon re-entry. Sean and I talked, got all our stuff moved out of the service module and Aquila that we needed and put it inside of Lil' Dipper. Then we started up Lil' Dipper with a very complicated set of maneuvers and got ourselves on course for our set target (Earth of course). We were going to bring up Matt's limp body with 12 hours till re-entry, so we were good there (we had at least 2 days till we re-entered anyway). Sean and I got in our sleeping bags and went ahead and logged out. We then got ready to sleep and prepared ourselves for tomorrow. This would be one of the last times I would sleep in space.

After Orion 9, I was scheduled to be on Orion 14, which would carry me to the Moon along with Sean. Even if Matt was still alive, he wouldn't command us on that mission however. Sean would be commander of Orion 14, I would be LEM (Lunar Excursion Module) pilot, and Martin Brown would've been pilot, however it was changed to be a new person, named Allan Plutonium. He would be a rookie to space travel, so we would be in for a rocky ride!

"Sean are you prepared for jettison of the service module soon?" I asked

"You bet I am! Make it special though. This is my first flight that I've had as an astronaut," he said.

"Sure thing, you'll see one similar for Orion 14 anyway! Hah hah," I said.

"Houston, we're standing by for S.M. jettison in 30 minutes," I said.

"Copy that Lil' Dipper. You're currently 1 day, 2 hours, 49 minutes, 54 seconds from re-entry," they responded. I floated through the service module and got Matt out and into the MEM pilot's seat. It had the least amount of switches and would be fine to have him there. I buttoned up the hatch once I got my crucial items, Sean's crucial items, Matt's (also) crucial items, and once Sean did a fly through of it to make sure. By this point our stuff was overflowing into Aquila, so some items would have to be dumped. I got strapped into position and prepared to detach from the service module that had done this loooonnngggg journey with us.

"Houston, we're jettisoning the service module in 3...2...1...jettison!" I yelled. I flipped the switch and heard the bang of the separation. I then jetted forward with RCS on Aquila and turned it around to get a good look at the service module.

"Oh my gosh….," I said under my breath. "Houston, we're getting our first look at the service module now. We're seeing the real damage currently. It appears as if the entire bottom of the spacecraft has been blown to bits and is radiating energy outward. It's destroyed the lab, exercise bay, licked at the side of the rooms, and….oh no…..," I reported to Houston.

"Houston, this is Sean. It's blasted all the way up to the decoupler on the heat shield, which means the heat shield is most likely radiating and crippled," Sean said.

Complete silence in Mission Control. Finally, they responded.

"Well, Aquila it's impossible to do a correction burn strong enough now. We could raise up your orbit and burn you guys into a nearly captured orbit, but not quite captured. It's either try your chance with that heat shield, or be flung off into space never to be seen again," Houston said.

"Roger that. We'll try our chance with the heat shield," Sean and I said.

The rest of the time in space on Orion 9 was rather boring. I never heard the voice again, and the Earth got awfully big in the window. No longer were we feeling glum about the mission. We actually thought that we could do this.

Once we were within 2 hours of re-entry, we decided to transfer the rocks and such over. We got the most important stuff in, and were prepared to jettison the lander, Aquila, which had served us so well.
"Houston, we're prepared to jettison the MEM! We'll do so in 30 minutes," I said.

"Roger that Reed. Everything seems good down here!" they said. We floated while we checked the batteries on Lil' Dipper. All of them except for Batt. A and B were good. We'd just have to go down on 3 batteries, which should be enough, since we need 2 for the heat shield, and 1 for the utilities during and after re-entry. The other 2 were just backup in case we happened to have a batt. drop out. We didn't have any margin for error though. If we lost just 1 battery, we'd be utterly screwed.

30 minutes later, we went ahead and jettisoned Aquila.

"Houston, we have jettison of Aquila in 5...4...3...2...1...jettison!" I said. Just then we heard a big bang as the docking ports cut away from each other, and then we heard some tiny retro-rockets propelling it away and causing it to be in a burn up course for Earth.

"Lil' Dipper, coming up on re-entry in 5 minutes," Houston said.

"Copy that, batts still holding what they were about 30 minutes ago," I said. We were heading home. I saw the Earth in the window. We were about 600 meters from the atmosphere.

"Lil' Dipper, we're going to lose communication with you in 3...2...1...," Houston said. We were in the atmosphere. Flames were licking at the side of Lil' Dipper. Sean was piloting us through the dense atmosphere, and making sure we weren't too steep or shallow. If we were either, we'd be goners. We had no communication with Houston. Nothing was working in the commander's seat. We saw alarms flashing at us, and my heart sunk. I told Sean to pitch up and around to get us facing toward the ground. We were flying through the atmosphere at over 5000 meters per second.

Chapter 12

"Sean, hold our position. We can get ourselves through at this rate," I said. I saw particles flying outside, and knew we were in trouble. The thing that was worse is that the heat shield alarm went on at the exact moment I saw the particle. Our heat shield was failing. The charge wasn't great enough, the rate of speed was too strong, and we had a crack presumably.

The heat shield was blazing, and lights were flashing all over the place. We were losing oxygen, our $Co2$ levels were dangerously high, and our batteries were failing.

"Sean, we have 3 minutes till expected touchdown!" I yelled

"Copy that, I see the same on my clock!" Sean hollered back. Our communication with Houston was lost. Nothing was online except for our parachute systems which were running off of emergency reentry batteries. Soon those would be drained too.

Everything was pointing toward us NOT returning safely. Explosions were heard in the command pod, many switches went offline, causing Sean to conk his head on the panel. Within seconds the pod was filled with the flashing of red lights of the switch he pressed. I decided to pilot us manually, because the auto control was causing us to flip around and start burning up even faster.

Back at Mission Control, they waited with eager expectations. Blackout of radio signal for interplanetary flights was only supposed to last for 5 minutes. Anything under, and we probably either broke up or it was sheer fluke. If we aren't back in 6 minutes, we'll be presumed dead. Already we had passed 4 minutes into our re-entry. Mission control held their breath as they waited for any sign of life, or of death, should we fail.

5 minutes passed, and they called for us.

"Lil' Dipper, this is Houston. Do you copy?" they asked. No reply. They tried again. Once again, no reply. We were half way through our time of expected signal. Nothing was heard for 30 more seconds.

"Ok flight, that's 6 minutes and counting," ECOM reported glumly. He knew that we were past our time of expected signal. They held their breath, and were sad. Suddenly, they heard a slight crackling, and saw an image of Lil' Dipper's chutes coming out.

"Houston, this is Lil' Dipper, reporting from home!" Sean said. Cheers were heard in Mission Control. We had successfully made it to Mars and home, with a crippled ship and a broken Martian Excursion Module. We had done it!

"Houston, we're currently splashed down and waiting for the helicopters to come and pick us up!" I said.

"Roger that Lil' Dipper, welcome home!" Houston said. We saw the helicopter people knock on the hatch and tell us to open up the door.

"Houston, this is Reed, the ship is secure and well. This is Orion 9 signing off," I said. Then I opened up the hatch and was welcomed back to a great scene. I then felt a jerking on my left arm, and found that Matthew was alive!

"Matthew! You're alive!!!!!" I yelled.

"Yeah, I don't know how I passed out for that long like that!" Matt said. "Houston, this is Matthew Psy, and now we're really signing off Orion 9 for good," he said to Mission Control.

I think Houston was as shocked as I was. They welcomed him back and appointed him commander for the rest of the flight. Sean left the ship first, as this wasn't his ship. I left the ship second. Then finally, Matthew Psy, commander of the worst survived disaster in space flight since Apollo 13, left his ship, Orion 9, for good. We watched the capsule, Lil' Dipper, go off into the distance for the last time.

Chapter 13

Our mission was called what the title of this journal of the 3 of us is called. The Journey of a Thousand Years. It's also what our landing site was called, and was a name I was willing to take. Our samples were given to Houston, and we did all of our de-briefings. Then the 3 of us went our ways and did our things.

Matthew Psy, the commander of Orion 9, was scheduled to go to the International Space Station for a long duration mission with Orion 13. Orion 13 was his last flight however, as he was in it when it re-entered. The command module, Zeus, roared through the atmosphere of Earth in 2032. It broke up, and all 4 astronauts on board died, including pilot, Matthew Psy.

We thankfully were able to recover the flight log of Orion 13 and Matt's journal through it, which had survived the rigors of reentry, splashdown, and the swim. We found that the last moments of Orion 13 were not pretty. The command module's control panel was ripped out from the sockets, leaving it uncontrollable except from manual. Matt pulled the manual control and started steering them. When he did, a window ripped apart, causing flames to enter the cockpit. One astronaut, the mission specialist 1, flew out of the window and had contact lost in under 0.01 seconds.

The g's were incredible Matt stated. It soared to over 10 g's, which is unhealthy. He was thrown forward in his belt, and what do you know? The window closest to him had broken free, causing him to barely stay inside. It widened to a huge hole, and Matt threw his book to the commander. He then said"It was a pleasure sir" and lept from the capsule. The second pilot and the commander were all who were left. He started to steer it into position to get them through the rigors of reentry, when all heck broke loose. The pod's floor blew, and the commander was sucked into the near vacuum at his altitude. The secondary pilot was flung through the docking hatch which blew open in the extreme g's. Sadly, no one survived the terror of the flight. It was later determined that a damaged coil inside the parachutes had sparked and caused an explosion which blew the control panel.

    Sean Argon, the Martian Excursion Module pilot for Orion 9, was scheduled to go on Orion 14 with me, Orion 19 (to Mars again, but as commander this time), and Orion 25, heading off to an asteroid. We flew Orion 14 off to the Moon, and landed safely. It was my first steps ever being taken on the Moon. We returned safely and had a good time after. Perhaps if I can find my darn journal for that flight, I can write it down and show it to all of you as well. Orion 19 was a successful flight, at least much better than ours was. It was his second time on the Martian surface. Orion 25 was to be his last flight, however it caught fire on the pad similar to Apollo 1. Two of the 4 astronauts died, however Sean was one of the lucky two.

    As for me. Well, I was the pilot of Orion 9, and also commander for quite a bit of the time. I flew as commander of Orion 14, as Lunar Excursion Module Pilot of Orion 16, as Martian Excursion Module Pilot of

Orion 19 (my first steps on Mars), as pilot of Orion 22, and on Orion 26. I took Sean's place as the first man to step on one.

Sean and I live very close to each other in California. We use to live in Houston, Texas obviously, but we are now retired. Our flights with Orion were complicated and difficult, each one presenting different challenges and difficulties, but in the end, it was an amazing time in the space program.

Also, remember that voice that I heard in the service module? Well, guess who it was? It was Sean pulling a prank on me. Apparently I was deaf because most of the time I can tell if someone is faking a voice. However, he got pretty scared with my response of it.

I have looked up at Mars lately, seeing it through a telescope. I've been wondering, as I should be. We've had 45 Orion's so far, and we're getting close to the retirement of what America's manned space flight program has been for over 20 years. It's been almost 16 years since those fateful words and moments of Orion 9. As a 54 year old now, I've lived many days and moons as a plain civilian. Sean and I are very similar to each other. We were on many of the same flights, we study the stars for a living, and we're even manufacturing Anti-Matter for NASA's flights to Alpha Centauri.

Sean and I have been developing ways to create space craft to bypass the speed of light. Einstein himself said that it was impossible to break the speed of light, however, we decided to put that to the test. We can accelerate atoms to NEARLY the speed of light. It just needs that extra boost to get it up to the speed of light.

Going the speed of light is extremely interesting. If you go the speed of light on a bike and someone is filming you, when you get off you'll actually be YOUNGER than the person who was filming you. It's rather interesting because that could allow us humans to live longer than nature intended. When we finally finish our research, we shall be done with our Astronomy and our work will be finished.

Chapter 14

It's been over 40 years since the launch of Orion 9 now, and I'm old now. I'm over 60, but to be honest, I forgot my actual age. Lately life has been rather depressing. Sean Argon, my friend for almost my entire life, died at the age of 64. He was at work at

NASA, building the last rocket of the Orion series. Orion 100 was to be launched soon, and it would be a trademark flight, bringing the honor, courage, pride, and hard work of the Orion program to a close. He died of a heart attack when preparing the SRB on Orion 100 to be tested upon firing. It was a tragic loss, but he died doing what he loved most.

Everything that we put into Orion has come out good. We've flown many missions to the Moon, Mars, asteroids, Phobos, Deimos, and have orbited Venus, Mercury (for one day, after that the command pod, Intrepid, couldn't handle the extreme heat), Jupiter, Mars and it's moons of course, the Moon, also of course, and even plotted charts to go to Pluto. Orion 99 left on it's historical mission to complete the S.A.S. (Space Atlantis Station). The ISS was deobrited about 4 years ago, and created a nice light show for us.

Orion 100 would make a historical journey, to visit the landing site of Apollo 11 and the landing site of OUR mission, Orion 9. Olympus Mons is predicted to be erupting for at least 100 more years, so it's a given to land there and see what we can see. I took Sean's place in assembling Orion 100, and I've been getting payed quite a bit.

Whenever I die, I know it will be one to remember. Neil Armstrong is a good role model to follow. He felt that he wasn't anything special, and that he was just the person who got the job done right. I've tried not to be boastful, but I think it was easier for me than it was for Sean. Sean had to deal with the fact that he was the second man on Mars and the only one alive who had been on Orion 9 and stepped on the surface. As for me, I orbited Mars on Orion 9, so I can't feel too special hah. Whatever the case is, I trust that my instincts will allow me to live these years of mine with a purpose of heart and a special feeling. I must believe in my strengths and allow myself to fly with a purpose.

      Whatever happens, I hope it's good. I now know what I must do, and that's to live my life as a person of Earth. I have fulfilled writing this journal for future generations and people. The Journey of a Thousand Years. It's an amazing title to have, don't you think? The Orion rocket has served us well, as I watch Orion 100 launch for the final liftoff of Orion.

      "...and Liftoff! Liftoff of Orion 100, demonstrating the pride and hard work of the Orion program! The rocket has cleared the tower for this amazing flight." This was the last thing I heard in the world, and I left the world peacefully, undisturbed, with all that I aquired in my life at my bedside. My life has complete, and I shall now start the real Journey of a Thousand Years.

www.ingramcontent.com/pod-product-compliance
Lightning Source LLC
Chambersburg PA
CBHW051816170526
45167CB00005B/2035